The Cook and Peary Expeditions: The Histo
over Who Reached the N

By Charles River I

Robert Peary's expedition near the North Pole

About Charles River Editors

Charles River Editors provides superior editing and original writing services across the digital publishing industry, with the expertise to create digital content for publishers across a vast range of subject matter. In addition to providing original digital content for third party publishers, we also republish civilization's greatest literary works, bringing them to new generations of readers via ebooks.

Sign up here to receive updates about free books as we publish them, and visit Our Kindle Author Page to browse today's free promotions and our most recently published Kindle titles.

Introduction

Peary

The Cook and Peary Expeditions

"I have stated my case, presented my proofs. As to the relative merits of my claim, and Mr Peary's, place the two records side by side. Compare them. I shall be satisfied with your decision." – Frederick Albert Cook

"Whatever the truth is, the situation is as wonderful as the Pole, and whatever they found there, those explorers, they have left there a story as great as a continent." - Lincoln Steffens

It is the dreamland of most children in Europe and the Americas, and the mysterious home of the mythical Santa Claus, his devoted wife Mrs. Claus, the reindeer and the many elves who make Christmas toys each year. In many ways, the North Pole is the first geographical location many kids learn, if only because children over the age of 3 can manage to tell any interested adult that Santa Claus lives there. In reality, of course, the North Pole proved to be as elusive for many brave explorers as jolly old Santa has been for children who wait up at night by the chimney.

The biggest problem, of course, is the North Pole's unforgiving location, far from sunshine or any sort of natural warmth. Another problem, one that would only became obvious in the 20th Century, was that it is located not on any piece of stable land but in the middle of the Arctic Ocean, usually covered by ever shifting ice floes. Finally, without modern technological advances, it was nearly impossible to tell when one has actually reached the planet's northernmost spot.

This has led to more than one argument about who actually made it and who did not; as historian E. Myles Standish put it, "Anyone who is acquainted with the facts and has any amount of logical reasoning can not avoid the conclusion that neither Cook, nor Peary, nor Byrd reached the North Pole; and they all knew it." Those sentiments were echoed by Canadian explorer Richard Weber, who asserted, "We came to the conclusion that Peary never got anywhere near the Pole. On the ice, everything looks the same. I'm afraid we'd have been lost without a global positioning system."

The controversy truly began on September 1, 1909, when the *New York Herald* printed a headline that told readers, "The North Pole is Discovered by Doctor Frederick A. Cook." By mid-1909, almost everyone in the polar establishment believed that Frederick Cook was dead, since his expedition had not been seen or heard of for a year. Then, suddenly, the *New York Herald* broke the news - the indestructible Cook had returned to civilization, and what's more, he had reached the North Pole. The newspapers hailed it as a great American achievement, and overnight Cook found himself a hero and a major celebrity.

However, less than a week later, on September 7, 1909, a rival newspaper, the *New York Times*, published their own version: "Peary Discovers the North Pole After Eight Trials in 23 Years."

Who was to be believed? The physical characteristics of the North Pole were known to none, so no viable comparisons could be made, and since the North Pole lay on a shifting continent of ice, its position might be in one place today and another tomorrow.

From the moment that ancient Greek thinkers determined the Earth was a sphere and gave it both an equator and an axis, the question of the nature of its poles commenced. By the late 19th century, it was understood, or at least strongly believed, that the North Pole did not lie on land but was an oceanic point. It was also believed that beyond a carapace of ice, lying more or less at 80° of latitude, there existed an open polar sea. The basis of this belief was simply that ice only formed in the proximity of land, and where there was no land, there could be no ice. The observations of whalers and various travelers and fur traders hinted even more strongly that a society of living things resided beyond the boundaries of known geography. And besides that, the perpetual sun of an Arctic summer would surely melt the ice, and if a ship could simply breach the ring of ice surrounding the North Pole, the open polar sea would be wide open to navigation. Beyond that may even lie the Northwest Passage.

In 1827, English maritime explorer William Edward Parry mounted the first serious attempt to reach the North Pole. From Verlegenhuken, Svalbard, Parry set off across the ice, hauling two small boats, in the belief that at some point open water would be found. It was not, and the effort was eventually abandoned. In 1871, the U.S. Polaris Expedition was thwarted by poor planning and mutiny, and five years later the British Arctic Expedition of 1875-1876, led by Sir George Strong Nares, was also turned back. A further American effort in 1879 was struck by disaster when its ship was crushed and sunk.

It was not until 1895, when the legendary Norwegian explorer Fridtjof Nansen set off across the ice on skis, that a successful navigation to the North Pole began to slip into reach, and there were other attempts before the end of the century. In 1897, for example, Swedish engineer Salomon August Andrée made an attempt to reach the North Pole in a hydrogen balloon. In 1900, Italian explorers Luigi Amedeo, Duke of the Abruzzi, and Captain Umberto Cagni, set off by dogsled and reached a latitude of 86° 34'. The North Pole itself, however, remained out of reach.

The Cook and Peary Expeditions: The History and Legacy of the Controversy over Who Reached the North Pole First chronicles the groundbreaking expeditions. Along with pictures of important people, places, and events, you will learn about the expeditions like never before.

The Cook and Peary Expeditions: The History and Legacy of the Controversy over Who Reached the North Pole First

About Charles River Editors

Introduction

   Previous Attempts to Find the North Pole

   The Protagonists

   The North Pole Controversy

   The Aftermath and the Investigations

   Online Resources

   Further Reading

Free Books by Charles River Editors

Discounted Books by Charles River Editors

Previous Attempts to Find the North Pole

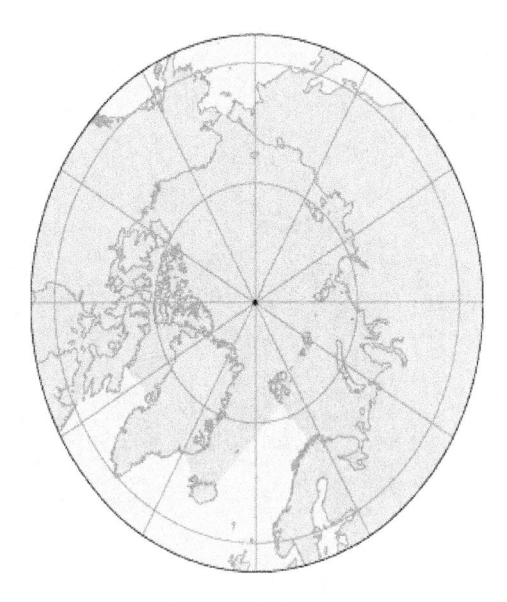

The location of the North Pole

Mercator's map and its location of the North Pole

"I very much rejoice to see that Arctic research is to be renewed by British explorers, and that the subject brought forward by Captain Shepard Osborn has been taken up by yourself and the Royal Geographical Society. Now that most of the mysteries of the interior of Africa and Australia have come to light, the greatest geographical problems that remain to be solved are the geography of the Central Polar Regions, and the attainment of the Poles themselves; and it is my conviction that the English nation, before all others, is destined, or at least is in the best position, to achieve this, the great crowning triumph of the discoveries on our planet." - Dr. Augustus Petemann, member of the Royal Geographical Society of London, 1864

It is very unlikely that anyone will ever know the name of the first person to reach the North Pole; in fact, it is unlikely that whoever that person was even knew what he had done. Most likely, he was a member of an Inuit or Viking crew blown off course by a terrible winter storm, and while it's possible that person or people lived to tell about the trip, it's highly unlikely. According to *The Saga of Ara Frode*, written sometime between 1067 and 1148, "The land which is called Greenland was discovered and settled from Iceland. Eric the Red was the name of the man from Breidafjord, who sailed thither from thence and there took land at the place which is since called Eiriksfjord. He gave the land a name and called it Greenland, and said that having a good name would entice men to go thither." Given the Viking activity around Greenland, an inadvertent trip to the North Pole was quite possible.

Ultimately, however, expeditions that neared the Pole would come centuries later, such as the one led by Sir Hugh Willoughby, who left England in 1553 in search of distant shores and hoped to reach Cathay through the Northwest Passage. "The courtiers came running out and the common people flocked together, standing very thick upon the shore; the Privy Council, they looked out at the windows of the court, and the rest ran up to the tops of the towers." Unfortunately, instead of making it to Cathay, Willoughby and his men got caught inside the Arctic Circle. Somewhere off the coast of Lapland he wrote, "We went into the harbor and anchored at a depth of six fathoms. This harbor juts into the mainland for about 2 miles, and the width is half a league. There were a lot of seals and other large fish, and on the mainland we saw bears, deer and other large strange animals and birds, such as wild swans, gulls, and other, unknown to us and excite our surprise. Having stayed in the harbor for a week and seeing what time of the year and later established that the weather is bad — from the cold, snow and hail, as if it was in the middle of winter, we decided to spend the winter here. Therefore, we sent three people to YU.-YU.-Z. to see if they do not find people, and they were three days, but people have not found, and after that we sent 4 more people to the west, but they returned without finding any people. Then we sent 3 people in the south-east that are returned in the same order, having found neither people nor any were home."

Though they had plenty of supplies and were able to create shelters for themselves, the men did not survive the winter. Instead, a group of Russian sailors found their bodies the following May, and one of them wrote, "Some of the dead were found sitting with pen in hand and paper in front of them, others — sitting at a table with plates in hand and spoon in his mouth, and others — opens the closet, others — in other positions, like statues, which have put in this way. Just looked like a dog."

An illustration depicting the deaths of Willoughby and his crew

A few decades later, in 1596, Dutch explorer Willem Barentsz led another expedition into the Far North and became the first European to make it to Spitsbergen, an island in what is now Northern Norway. There they observed the phenomenon of the long Arctic night, and according to one scholar who interviewed the men when they came home, "they lost the Sun on 4 November of the year 1596 and saw the Sun again on 24 January of the year 1597 at the same latitude of 76 degrees, where they had constructed their house on Nova Zembla. ... Because it had been always day without darkness for them, for longer than ten weeks, and because during that same period of time the skies had not been clear enough to count the revolutions of the Sun, I asked them how they knew that it had been accurately on the 4th of November that they lost the Sun...they answered me that they had always had their clocks and sandglasses ready, so that they had been certain to have the time right."

Willem Barentsz

In 1610, increased interest in Arctic exploration was ignited when Jonas Poole set out, as he said, to "catch a whale or two." However, he did not plan to visit the overfished shores already in use but to go much further, having been given the following commission by his sponsors: "Inasmuch as it hath pleased Almighty God, through the industry of yourself and others, to discover unto our nation a land lying in eighty degrees toward the North Pole: We are desirous, not only to discover farther to the northward, along the said land, to find whether the same be an island or a main, and which way the same doth trend, either to the eastward or to the westward of the pole; as also whether the same be inhabited by any people, or whether there be an open sea farther northward then hath been already discovered: For accomplishing of all which our desires we have made choice of you, and to that end have entertained you into our service for certain years upon a stipend certain; not doubting but you will so carry yourself in the business for which you were so entertained, as God may be glorified, our country benefited, yourself credited, and we in our desires satisfied, etc."

While he may not have lived up to all these lofty goals, Poole did learn a great deal about the western coast of Spitsbergen and saw a "great store of whales," igniting a frenzy of English whalers to head to the Arctic. By 1613, ships from four different European countries were competing with each other for whales off Sritsbergen's coast. Explorers continued to make their way further and further north until, in 1648, two Russians, Semyon Dezhnyov and Fedot Alekseyev, made it through what is now known as the Bering Strait.

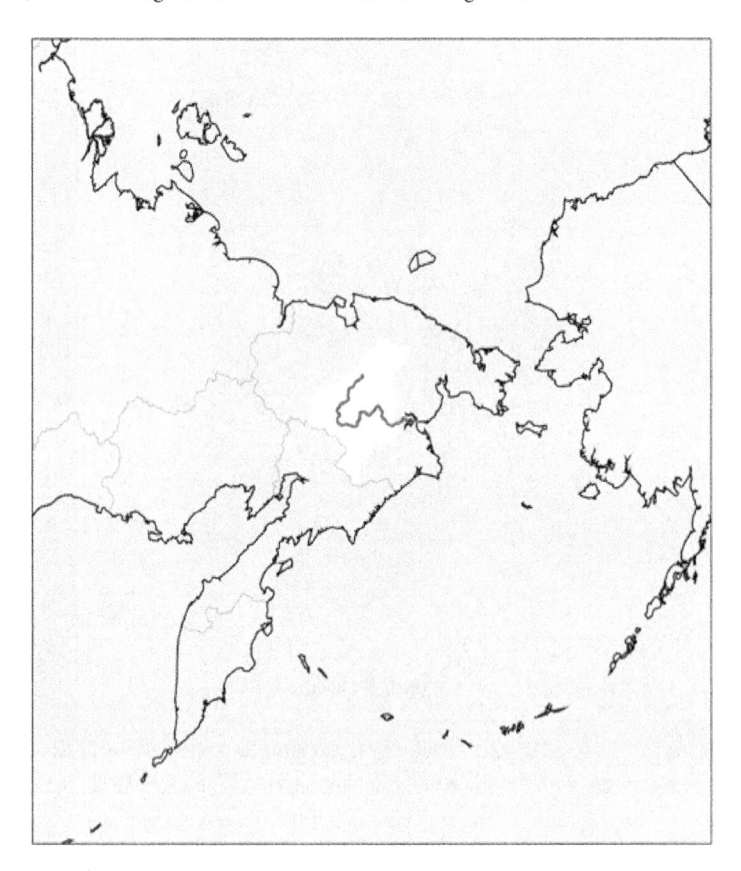

Map of the Bering Strait between Russia and Alaska

Next came the Great Northern Expedition, which left Russia in 1733 and lasted 10 years, during which members mapped most of the northern coast of Siberia. Led by Vitus Bering, who later lent his name to the Bering Sea, the Bering Strait and the Bering Land Bridge, it was one of the largest expeditions in history. Finally, in 1773, a young Horatio Nelson was among the crew of the HMS *Racehorse*, which, along with the HMS *Carcass*, reached 80° 37' N, a new record in travel into the Arctic Circle. When the ships found themselves stuck for a time among a number of ice floes, young Nelson, then only in his teens, set off to hunt a polar bear. However, when he was close enough to shoot, his musket misfired and he found himself face-to-face with a frightened animal. According to legend, Nelson had to fight the bear off with the butt of his gun until the ice between them broke and they floated apart. Seeing what was happening, the captain also ordered a cannon fired, which scared the bear away. When the captain then had Nelson rescued for the express purpose of berating him, Nelson justified his little adventure by saying he had hoped to take the bear's skin home to his father.

Bering

Nelson

The early 19th century saw much of Europe embroiled in the Napoleonic Wars, but the climactic end of the era after the Battle of Waterloo in 1815 brought about the beginning of a new time of prosperity and exploration, especially among the triumphant British. In 1827, William Edward Parry, a British naval officer and veteran of the wars, set sail for the arctic, determined to reach the North Pole. In this sense he was a pioneer, for he was one of the first to plan an expedition with that specific goal in mind. His plan as a simple one. Sail as far north as he could and then, when he could sail no further because of ice, drag small whaling boats across the ice to the next place where there was water. According to his memoir, written by his son, "at five p.m. on the 21st of June, the two boats, 'Enterprise' and 'Endeavour,' respectively commanded by Parry, and his lieutenant, James C. Ross…set out for the northward. The boats employed on this novel service were fitted with strong 'runners,' shod with smooth steel, in the manner of a sledge, to the forepart of which the ropes for dragging the boat were attached. … For three days they sailed through the open water, but the ice gradually gathered round them, until, at length, they were compelled to haul the boats up on to the floe, and the actual labor of the expedition now commenced. Unless compelled by circumstances to act otherwise, the usual plan was to travel only by night, when the snow was harder than during the day time. It will, however, be remembered that the daylight was constant, and that the terms ' day ' and ' night ' were only relative…. The rough nature of the ice, combined with the softness of its upper surface, rendered each day's work very tedious and laborious. Often, their way lay over small loose rugged masses, separated by little pools of water, obliging them constantly to launch and haul up the boats, each of which operations required them to be unloaded, and occupied nearly a quarter of an hour."

Parry

Unfortunately, even while they were sailing their boats north, the ice on which they were traveling was slowing floating south, which meant much of their efforts were in vain. "For a few days more they persevered, in the face of heavy snow-storms, and torrents of rain, which Parry had never seen equaled, but, the drift of the ice continuing as great as ever, he was, at length, compelled to confess that further labor were useless. Between the 22nd and 26th of July, they had dragged the boats only ten or twelve miles, and were, after all, actually three miles southward of the point they had reached on the first of these days. … One day's rest was given, for the men to wash and mend their clothes, while the officers occupied themselves in making observations in the highest latitude which had ever been reached, viz. 82° 40' 23". At this extreme point of their journey, their distance from the *'Hecla,' after five weeks travel, was only 172 miles, to accomplish which they had traversed upwards of 290 miles with the boats."

In fact, the crew had reached 82° 45', a new record north, but they hadn't reached the North Pole itself, and for nearly 50 years, no other efforts were made. Then, in June 1871, an American named Charles Francis Hall set out from New York City in a ship called the *Polaris* in hopes of reaching the North Pole. The expedition was plagued by disaster, some of it apparently manmade; Hall himself died suddenly on November 8 in Greenland, apparently the result of arsenic poisoning. Without his leadership, the expedition broke up at one point, with 19 crewmen becoming trapped on an ice floe, where they survived for six months before being rescued. Meanwhile, the *Polaris* itself ran aground near Etah, Greenland in October 1872. In spite of all these setbacks, according to the Geographical Society of Paris, "the Polaris…passed beyond Smith's Sound and Kennedy Channel, as far as 82° 16' — that is to say, the nearest to the pole that any vessel has reached under sail…More than 700 miles of coastline have been discovered and reconnoitered."

An engraving of Hall

Others continued to make efforts in the direction of the North Pole, with the British Arctic Expedition reaching 83°20'26" in May 1876. According to Commander Albert H. Markham, in

May 12, 1876, "[t]he remainder of the party carrying the sextant and artificial horizon, and also the sledge, banners and colors, started northwards. We had some very severe walking, struggling through snow up to our waists, over or through which the labor of dragging a sledge would be interminable, and, occasionally, almost disappearing through cracks and fissures, until twenty minutes to noon, when a halt was called. The artificial horizon was then set up, and the flags and banners displayed. These fluttered out bravely before a fresh S.W. wind, which latter was, however, decidedly cold and unpleasant. At noon we obtained a good altitude, and proclaimed our latitude to be 83° 20' 26" N., exactly 399^ miles from the North Pole. On this being duly announced, three cheers were given, with one more for Captain Nares ; then the whole party, in the exuberance of their spirits at having reached their turning point, sang the 'Union Jack of Old England,' the 'Grand Palaiocrystic Sledging Chorus,' winding up, like loyal subjects, with 'God save the Queen.'…The instruments were then packed, the colors furled, and our steps retraced to the camp."

The final American attempt of the 19th century began in 1879 but ended in disaster when the ship the group was traveling on, the USS *Jeanette*, was crushed by an ice floe. George W. DeLong, the commander of the expedition, was killed, along with more than half his crew. Meanwhile, the next advance toward the North Pole was made in April 1895 by Fridtjof Nansen and Hjalmar Johansen of Norway. Sailing on the *Fram*, they spent more than two years trying to make it to the elusive pole, and Nansen later summed up their journey: "Tuesday, May 19, 1896. We were frozen in north of Kotelnoi at about 78° 43' north latitude, September 22, 1893. Drifted northwestward during the following year, as we had expected to do. Johansen and I left the Fram, March 14, 1895, at about 84° 4' north latitude and 103° east longitude, to push on northward. The command of the remainder of the expedition was transferred to Sverdrup. Found no land northward. On April 6, 1895, we had to turn back at 86° 14' north latitude and about 95° east longitude, the ice having become impassable. Shaped our course for Cape Fligely; but our watches having stopped, we did not know our longitude with certainty, and arrived on August 6, 1895, at four glacier-covered islands to the north of this line of islands, at about 81° 30' north latitude, and about 7° E. of this place. Reached this place August 26, 1895, and thought it safest to winter here. Lived on bear's flesh. Are starting to-day southwestward along the land, intending to cross over to Spitzbergen at the nearest point."

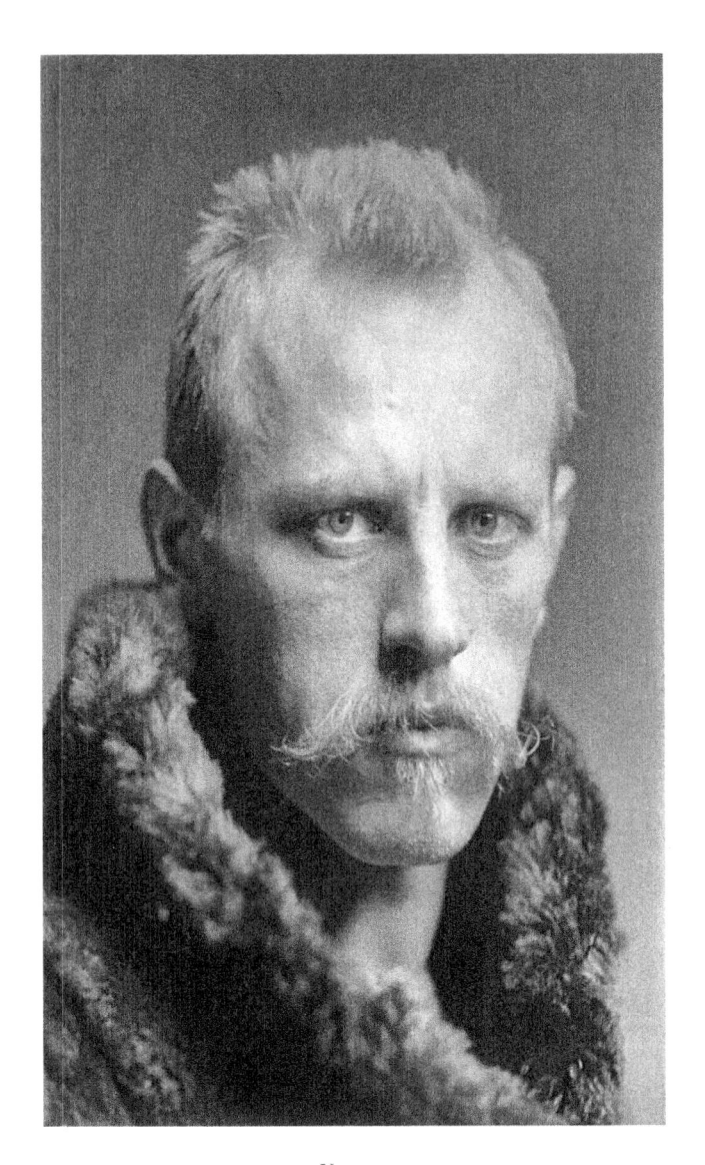

Nansen

Nansen's ship in the Arctic ice

Nansen's record stood until 1900, when, on May 11 of that year, Captain Umberto Cagni of the Italian Royal Navy reached 86°34'. Unaware that this point was as far as he would make it before having to turn around, Cagni wrote in his diary, "This morning we were so warm that we perspired even when standing still; at eleven, after taking a light soup, we get into the sleeping-bag, where we sleep with half our bodies outside, without even making use of the little bags which serve as pillows. The temperature is -7. At mid-day I take a tolerably exact meridian altitude, and in the evening am able to take an excellent observation of the horizontal angle, which gives me for longitude 50 1.5V and we are in 82 34' N. lat. I shall go on thus for a few days more, and then, if I find that I cannot overcome this drift, I shall push on resolutely to the south, as far, if necessary, as 81 30', in order to take shelter among the islands, as I am convinced that the powerful current which causes me so much anxiety exercises no influence in the extent of sea comprised between them and the line which joins Cape Fligely with the western extremity of Alexandra Land. In this case we shall have to re-ascend towards the north along the land, which will lengthen our journey very much, but I do not see any means of escaping."

Cagni

The Protagonists

"Men Wanted for hazardous journey. Small wages, bitter cold, long months of complete darkness, constant danger, safe return doubtful. Honor and recognition in case of success." – Ernest Shackleton

Frederick Albert Cook was born on June 10, 1865 in Hortonville, Sullivan County, New York, the son of a German immigrant physician, Doctor Theodore A. Koch. His father died when he was a child, and his mother raised the family alone on the hardscrabble streets of New York. Frederick decided he would follow his father into medicine, and he enrolled at the New York University medical school. He ran a business delivering milk to pay for tuition, and entertained himself on books and publications covering the business of exploration.

Cook

He is described in most of his biographies as an "ordinary looking man," with blue eyes, standing 5'9 and weighing around 175 pounds. His shoulders were wide, and his chest deep.[1] He emerges from his own autobiography as a more mystical, restless, and discontented youth, constrained by poverty and dreaming of a larger, grander life. He read books and newspapers, and he promised himself that someday he too would be a member of that fraternity of heroes.

While still at medical school, he married, and then in 1890, he graduated and opened a medical office in Manhattan. At some point in the months that followed, however, his wife died of peritonitis in childbirth. He consoled himself by reading voraciously of travel and exploration, looking around for such an opportunity.

That opportunity came when he opened his morning newspaper and was attracted to an article covering the proposed expedition of a certain U.S. Navy Lieutenant, Robert Edwin Peary. Peary

---
[1] Eames, Hugh. *Winner Lose All: Dr. Cook and the theft of the North Pole*. (Little, Brown, London, 1973) p5

was at that time stationed in Philadelphia, and he was preparing an Arctic expedition for which he required a surgeon. Cook immediately submitted an application, and two months later he was surprised and gratified to receive a summons to Philadelphia to meet Peary for an interview. Cook impressed Peary, and on the surface they seemed to like one another. Cook, however, was warned, "The life up there under the Pole is terribly hard…we will be as much out of touch with the world as we would on some other planet. Some of us more than likely will never return…'[2]

Cook needed no persuasion. Still grieving for his wife and child, it was an opportunity to distract his mind, but more importantly, it offered an opportunity to be peers with men who he admired. Peary was certainly such a man, and by the end of the day, he was signed up as the first official member of Peary's expedition. His contract specified that he would receive no pay.

Robert Peary's biographical record is surprisingly thin. Although identified with the Arctic, Peary, in reality, used Arctic exploration and the potentialities that it offered solely as a vehicle to acquire a personal reputation and fame. That he was a robust and competent man goes without saying, but unlike many others in the modern field of exploration, he had very little authentic interest in the Arctic, and certainly no sincere interest in its people.

Peary

---
[2] Quoted: Henderson, Bruce. *True North: Peary, Cook, and the Race to the Pole.* (Norton & Co. New York 2005) p41

In numerous letters to his mother, endlessly quoted in his many biographies, Peary repeats time and again, without any effort to disguise his objectives, that he desired above all else to be famous. He comes across at the very least as a bombastic and attention-seeking individual, self-serving to the point of narcissism, shrewd, and mostly unscrupulous.

When he and Cook met, Peary was 34 and his career was at a standstill. Born on May 6, 1856 in Cresson, Pennsylvania, he grew up in Portland, Maine, and graduated from Bowdoin College as a member of the Delta Kappa Epsilon and Phi Beta Kappa fraternities. He was a reckless and gregarious youth, tough and daring. His first employment was with the Coast and Geodetic Survey before he moved on to the U.S. Navy as a civil engineer holding the simulated rank of Lieutenant. He served primarily on the survey team studying the Nicaraguan isthmus as a possible site for the canal later built in Panama.

Somewhat later, while on furlough in Washington, D.C., he came across a pamphlet describing a recent Swedish expedition attempting to cross the Greenland ice cap, and an idea was sown in his mind. The Arctic remained the Holy Grail of geographic exploration, the last great unclaimed wilderness, and this, he recognized immediately, would bring him the fame he so desired. Thus, in the summer of 1886, he sailed north with a three month accumulated leave, and $500 borrowed from his mother. It was a haphazard expedition, but nonetheless, in May 1886, he left Sydney, Nova Scotia, and struck north. He made a credible effort, but very quickly his amateurism and lack of experience began to tell, and he was turned back by a combination of weather conditions and supply difficulties.

Despite the failure, this gave him a foot in the door when it came to the Arctic exploration community, and on the basis of that, he set about planning a second expedition. He also wrote a letter to his mother, articulating in his most famous quote precisely what motivated him: "My last trip brought my name before the world; my next will give me a standing in the world....I will be foremost in the highest circles in the capital, and make powerful friends with whom I can shape my future instead of letting it come as it will....Remember, mother, I must have fame."

As a result, Peary and Cook were men of diametrically different tempers and worldviews. Peary has been described as one of the last of the great "imperial" explorers, even though the United States was not then among the imperial nations. The objective of exploration during this age was to establish a Western nation's claims to territory, for the complex reasons of prestige, strategic location, and wealth. The methods of achievement were largely irrelevant, and such finer concepts as respecting and attempting to understand the alien cultures affected by the relentless advances had not yet matured.

For example, during an 1897 expedition to Greenland, Peary ordered his men to loot native graves for remains that were later sold to the American Museum of Natural History in New York City as anthropological specimens. He also returned with two Inuit men, one woman and three young children, and he presented them for study at the same museum. Within a year, four were dead from a strain of influenza to which they had no resistance. It was regrettable but justified. Natives in general were regarded as archaeological specimens, and they were treated as such.

Cook, however, came from a different tradition. His role on Peary's expedition was more than simply surgeon, but also "ethnologist," a discipline which he approached as an amateur, but nonetheless with an authentic and enthusiastic interest. His own later travels would often be in the company of the Inuit, and in time, he learned their dialects, adopted their diets and utilized their methods of survival. This was the birth of a new type of wilderness travel, less Eurocentric in tone and more sympathetic to vulnerabilities of natives. Furthermore, he was open to what they had to teach.

Needless to say, despite early prognostications of one another's excellence, largely for the sake of press releases, Cook and Peary did not get along at all. The expedition sailed out of Brooklyn on June 16, 1891, depositing the members at McCormick Bay on the west coast of Greenland, where it would thereafter be based. Peary succeeded in reaching latitude 82°, and in reaching that, he claimed to have reached the northern limit of Greenland. Peary, of course, guarded his right to publish the definitive account of the expedition, but in doing so he invoked a legal injunction against Cook, who wished to write and publish the results of his own ethnographic studies. This sowed the first bad blood between the two, and Cook would subsequently refuse to join further expeditions with Peary.

Ironically, Peary got lost in the Arctic in 1901, and his family approached Cook for help. Cook sailed north on a rescue ship, found Peary, and treated him for ailments ranging from scurvy to arrhythmia. The rift between the two men, however, was not healed by this; Cook remained outside of Peary's circle, which seemed to affect Peary as a sentimentalist more than it affected Cook.

While Peary's character was certainly questionable, Cook's track record was hardly spotless either. Sometime late in 1902, he happened to read a newspaper article about Mount McKinley, the great Alaskan mountain and the highest peak in North America. It had been "discovered" only five years earlier, and it had never been climbed. Here was an untapped opportunity to be the first person to do something, and though Cook had no particular experience in mountaineering, he still decided that he would climb Mount McKinley.

Almost entirely self-financed, his expedition got underway on June 25, 1903. In the company of five men and 14 pack horses, Cook traveled 500 miles overland before reaching the foothills of the great mountain. Traversing the southwest ridge, Cook achieved an altitude of 11,200 feet before running up against impassable cliffs, and with winter closing in, he was reluctantly forced to abandon the attempt. He returned to New York with a respectable weight of geological and botanical material, and in general the expedition was regarded as a respectable attempt to summit a great mountain.

Cook suffered no criticism for failing, and he had no reason to criticize his own effort. In 1906, however, he returned to Alaska to mount a second attempt on Mount McKinley. This time, with the benefit of his previous experience and some knowledge of the topography of the mountain, he felt himself better prepared. Included in this expedition were Professor Herschel Parker of Columbia University and the artist Belmore Brown. This created a certain amount of ballast, and with it came expectations. In all likelihood, a third attempt would never happen.

Parker

After two months of exploring and mapping the southern foothills of Mount McKinley, it was generally concluded by all of the experts present that none of the southern glaciers offered a viable path to the summit. Nevertheless, Cook was determined that he would try, and while climbing only in the company of a horse-packer by the name of Edward Barrill, he left the group and made what he described as a "successful" climb up Mount McKinley.

When Cook returned to his base camp and reported this fact, others were immediately suspicious. Brown calculated that Cook and Barrill had barely enough time to reach the base of the mountain, let alone the summit. In Fairbanks, and further south, bars, clubs and parlors rang with skepticism. There was no possible way that the New York doctor had reached the summit, and men who had mined the creeks of the interior for 30 years wrote it off as impossible. Belmore and Brown agreed.

Once back in civilization, they published their arguments and suspicions, but, relying somewhat on his established reputation as an explorer, Cook was able to press his claim, and the matter of his first ascent of McKinley was accepted as fact. The May 1907 issue of *Harpers Monthly Magazine* carried the first photographic images of Cook's ascent, republished later in Cook's own book of the expedition, *To the Top of the Continent*. Cook's main proof of his summit success was a photograph taken, ostensibly of the summit, but not from the summit, and none of his photographic records show any sort of view from the highest point. The figure included in Cook's summit image, presumably himself, appears obviously fake. Subsequent research by modern historians and mountaineers, studying Cook's photographs and juxtaposing them against known points, confirmed that none were taken at points close to the summit. None, in fact, were taken past a point known as the Gateway, at the north end of the Great Gorge.

Cook's staged photograph of the summit

When these images were examined by Brown and Parker, both men were convinced that the celebrated summit shot was a fake. It was a difficult point to prove, however, because soon afterwards Cook disappeared into the Arctic, only to emerge in September 1909 proclaiming himself to be the first to reach the North Pole. His reputation was soaring, and any challenge to it at that point would have seemed peevish and uncharitable.

Nonetheless, Parker and Brown, determined to expose Cook as a fraud, and, supported by the New York based Explorers Club (of which Cook was a founding member), they mounted their own expedition to further explore the Mount McKinley region. Both this party and another from Oregon attempted to duplicate the route described by Cook, but both quickly became bogged down on the impossible south side of the range. Other efforts on the north side were more successful, but what seemed to be proven was that Cook could not have realistically reached the summit of Mount McKinley from the south.Late in 1909, Edward Barrill signed a legal affidavit acknowledging that neither he nor Cook ever reached the summit, but it was later ascertained that he had been paid to issue this statement by Robert Peary in the midst of the controversy over who reached the North Pole.

With hindsight, it's clear Cook did not summit Mount McKinley, and his claim to have done so was nothing less than a fraud. Without an archive of prior ascents to compare it against, it was nonetheless accepted by the exploration fraternity and recorded as fact, but in hindsight, it's clear both explorers were willing to be unscrupulous when it came to making a claim to fame.

The North Pole Controversy

"[Cook] genuinely loved and hungered for the real meat of exploration-mapping new routes and shorelines, learning and adapting to the survival techniques of the Eskimos, advancing his own knowledge-and that of the world-for its own sake." – Robert Bryce

Cook's purported success in summiting Mount McKinley positioned him well for his first attempt to reach the North Pole. He lectured widely to an appreciative audience, and his break came early in 1907 when he made the acquaintance of a wealthy casino owner named John Bradley. The two planned a hunting and exploration expedition to Greenland, and with Bradley's financial backing, a fishing schooner was purchased, refitted, and renamed the *John R. Bradley*. The expedition did not specifically embark with the intention of attempting the North Pole, but Cook certainly had that goal in the back of his mind.

The expedition departed Gloucester, Massachusetts in the summer of 1907, heading through Baffin Bay to northern Greenland. There, at a native settlement known as Annoatok, a base camp was established, and at some point here, Cook revealed his intention to reach the North Pole. By then, however, it was late in the season, and the *John R. Bradley* needed to head south or run the risk of being iced in for the winter. Arrangements were made, the two men parted company, and by the seat of his pants, Cook mounted his expedition. In February 1908, with a party of 9 natives and 11 light sledges hauled by 103 dogs, the expedition departed. His proposed route followed the one described by Otto Sverdrup, the leader of a previous Norwegian mapping party.

According to the route described in his book, *My Attainment of the Pole*, Cook's party followed the musk ox breeding grounds observed by Sverdrup, traveling across the Ellesmere and Axel Heiberg Islands to Cape Stallworthy, on the northern edge of Axel Heiberg Island, and at the edge of the frozen Arctic Sea. This gave the expedition the advantage of accumulating fresh meat from musk ox, which allowed them to conserve their stocks of preserved pemmican. The expedition was not by any means lavishly supplied or equipped, and it could not have proceeded far without the help of the local Inuit. Cook had by then established a style of travel that was light and supported by local knowledge. His companions, of course, were mainly Inuit, and it was they who made it possible to live off the land. As Cook put it, "No group of white men could possibly match the Eskimos in their native element."

The average load of Cook's sledges was 550 pounds, comprising mostly food for dogs and men. He traveled with one European, Rudolf Franke, but Franke was soon left behind, ostensibly to guard a cache of furs. It's possible Cook did not wish to have the facts of his journey too closely observed, and perhaps he did not want an educated presence when the moment of discovering the North Pole was celebrated. Either way, his claim was to have reached the Pole after traveling north from Axel Heiberg Island in the company of two Inuit men, Ahpellah and Etukishook.

The journey Cook described in his book, *My Attainment of the Pole*, is a harrowing one north across the open ice. Cook described days of powerful gales that made every breath and every step a nightmare. He nonetheless reached a point that he determined was the North Pole on April 21, 1908. He took sextant readings, and took note of the topography around him, expressing his belief that they had arrived "at a spot which was as near as possible" to the North Pole. The party remained at that point for two days, during which time Cook added to his observations, confirming what he believed to be his position. Before leaving, he left a letter in a brass canister at the site to make note of his discovery.

A photo Cook claimed was taken at or near the North Pole

The story of Cook's return journey is among the epics of travel and survival. Cook was the first to observe a westward drift of the ice shelf that carried his group away from cached supplies and expected routes. Hungry and desperate, he and his two companions stumbled into Annoatok in April 1909 after some 14 months of Arctic travel.

In the controversy that followed, where Cook was and what he was doing for much of this time has been speculated over by historians about as much as his actual claim to have reached the North Pole. He claimed to have been diverted from his intended route back to Annoatok by encountering open water, which forced him onto Ellesmere Island, and as far south on the archipelago as Jones Sound, between the Ellesmere and Devon Islands. Living off the land and hunting game for survival, he spent part of the winter on Devon Island, eventually crossing the Nares Strait back to the mainland of Greenland, returning to his point of departure almost a year after he left the Pole.

The timings here are certainly interesting. He left Annoatok early in February 1908 and reached the North Pole about three months later. Then, from the North Pole, he was absent for a year before returning to Annoatok in April 1909. His own claim to have traveled over the Arctic islands is the source of much controversy, with some historians suggesting that the claim was legitimate, and others making an interesting observation. Cook's Inuit companions, Ahpellah and Etukishook, were located by investigators later attempting to piece together Cook's journey, and they sketched a route of their journey south that bore striking similarities to that taken by the fictional shipwrecked explorers in Jules Verne's novel *The Adventures of Captain Hatteras*. For example, the route the two Inuit traced on a map goes over both the "North Pole of the Cold" and the wintering site of the fictional expedition.[3] Both expeditions also visited the same area of Jones Sound in the hope of finding a whaling ship to return them to civilization. The implications of this are that Cook fictionalized elements of his return journey, leaving the question of his true whereabouts during this period unknown. What is certain is that he returned to Annoatok in desperate straits, threadbare and hungry.

It is at this point that the story becomes even more opaque. At Annoatok, Cook met an American sport hunter by the name of Harry Whitney, who told him that he was widely thought to be dead. Moreover, Whitney informed Cook that 8 months earlier, Peary had set off from Annoatok on his own expedition to reach the North Pole. Knowing Peary as he did, Cook realized that Peary would be going public with his own claim, which meant time was of the essence. He swore Harry Whitney to silence until he had an opportunity to make his own announcement. Then, having rested and recovered his strength a little, he set off on another punishing overland trek to the Danish trading post of Upernavik, 700 miles away. From there, he hoped to catch a ship to Copenhagen, and then another to New York City.

Whitney tried to persuade him to wait for ship traffic to resume in Baffin Bay, when he could return on a chartered vessel that Whitney had ordered for the following spring. That would take too long, however, so Cook placed in Whitney's care all of his instruments, as well as trunks containing detailed logs of the expedition. These he could not safely carry back to civilization on a dog sled, and once the announcement of his successful attainment of the North Pole had been made, he would be able to back up his claims with the necessary documentation, which would by then be back in his possession.

Peary had set off from New York City on July 6, 1908, precisely a year after Cook had departed Massachusetts. He traveled aboard the SS *Roosevelt*, under the command of Captain Robert Bartlett. The expedition wintered near Cape Sheridan on Ellesmere Island, and at the end of February 1909, they left Ellesmere Island, traveling north towards the Arctic Ocean. Peary traveled with a much larger expedition than Cook, comprising some 50 men, a similar number of sleds, and 246 dogs. He also employed a much more traditional system of travel, utilizing bulk and manpower to deposit caches en route in order to ensure a steady supply of food for the return journey. This system, as it had before, revealed itself to be top-heavy and unwieldy, and Peary found himself progressing at a daily average of only 13 miles.

---

[3] The *"Pole of the Cold"* described by Jules Verne was vaguely the point of coldest temperature not related to the North Pole.

The *Roosevelt*

At a point judged to be some 135 miles from the North Pole, Peary also made the decision to send back all but a handful of native support crew and Mathew Henson, his black companion who was a member of most of his expeditions. This, no doubt, was intended to obscure any potential difficulties or inconsistencies concerning his claim, and also, of course, to avoid sharing the laurels.[4]

---

[4] Matthew Alexander Henson was the first African-American Arctic explorer. He was recruited by Peary for an early expedition, and ended up accompanying Peary on seven voyages over a period of nearly 23 years. Henson typically served as a navigator and craftsman, trading with Inuit and learning their language. He was known as Peary's "first man," and his first choice as crew on all of his expeditions.

Henson

On April 6, 1909, Peary and Henson arrived at a place that Peary "felt" was the North Pole. When asked by Henson if they were at the North Pole, Peary replied, at least according to Henson, "I do not suppose that we can swear that we are exactly at the North Pole." Nonetheless, the flag of Old Glory was hoisted and the moment was solemnized. According to Henson's account, from which the story is largely drawn, Peary took sextant readings over the days that followed, after which they moved on, leaving a flag and a note in a tin.

Meanwhile, before leaving Annoatok, Cook carefully boxed up all of his expedition records, except for his diary. This included his sextant, compass, barometer and thermometer, all of which he left with Harry Whitney upon a firm promise that they would be returned to him in New York in the summer. He left Annoatok in the third week of April 1909, and a month later, after traversing the frozen west coast of Greenland, he arrived at Upernavik, where he made his announcement.

Whitney

News spread like wildfire. A few weeks later, on board the *Hans Egede* bound for Copenhagen, the rough, travel-worn Cook, his hair long and his clothes ragged, was a hero. On September 1, 1909, the *Hans Egede* called in at Lerwick, in the Shetland Islands, and there Cook cabled the *New York Herald* with the simple news that he had reached the North Pole on April 21, 1908.

Back on Annoatok, Harry Whitney was cooling his heels waiting for a boat to take him back to civilization. His chartered ship did not arrive, and it would be August before a ship would call in at the settlement. That ship was the SS *Roosevelt*, Peary's expedition ship, sent by his sponsors to collect him at the end of his expedition. At some point before then, Peary himself appeared out of the Arctic, somewhat less travel-worn than Cook, but no less ebullient, claiming that he too had reached the North Pole. When news of Cook's visit to the settlement a month or so earlier reached Peary, he discovered there were rumors among the Inuit that Cook had reached the North Pole. Whitney was interrogated, but he would reveal nothing more than the fact that Cook had returned to civilization safely. Peary, of course, assumed Cook was about to lay claim to finding the North Pole.

Then came a fateful turn of events. As Whitney was embarking the SS *Roosevelt* for the journey home, Peary observed the boxes and trunks being loaded on his behalf. He was told that these were Cook's instruments and records, which Harry Whitney had promised to return to him. Peary refused to let anything of Cook's come aboard his expedition ship, and there was nothing that Whitney could do about this. Thus, according to Whitney, he returned Cook's notes and equipment to shore and hid them among the rocks. A few days later, the SS *Roosevelt* set sail southwards.

None of Cook's equipment or records would ever be seen again. Harry Whitney's claim to have hidden them on shore is probably true, but he left no record of exactly where, and the possibility that they could have been later visited and destroyed cannot be discounted. Either way, all the technical records of Cook's expedition were lost.

On August 26, the SS *Roosevelt* called in at Cape York, on the northeast coast of Greenland, and there Peary discovered the truth. Cook was indeed en route south to declare he had reached the North Pole. When he heard, Peary was enraged, stating to anyone who listened that he would soon put Cook in his place. He ordered the captain of the SS *Roosevelt* to stoke her boilers and chart a course to Indian Harbor, Labrador, some 1,500 miles distant, the nearest possible telegraph station. Two weeks later, his ship dropped anchor and Peary hurried on shore. There, he telegraphed the *New York Times*, who owned the rights to the subsequent story: "Stars and Stripes nailed to the North Pole."

A few days before the news broke of Peary's successful journey to the North Pole, the *New York Herald* had already run a front-page story under the headline, *"Fighting Famine and Ice, the Courageous Explorer Reaches the Great Goal."* That explorer was Cook, and the Great Goal, of course, was the North Pole. The news was greeted by the general public with amazement and acclaim, and no doubt whatsoever was expressed in regards to the authenticity of the claim.

This however, began to change when, a few days later, the *New York Times* ran its coverage of Peary's return to civilization and his claim to the North Pole. The *New York Times* headline was a little more muted, but implicit was its understanding that Peary's claim was the only believable one. "Commander Robert E. Peary, U.S.N., has discovered the north pole. Following the report of Dr. F. A. Cook that he had reached the top of the world comes the certain announcement from Mr. Peary, the hero of eight polar expeditions, covering a period of twenty-three years, that at last his ambition has been realized, and from all over the world comes full acknowledgment of Peary's feet [sic] and congratulations on his success."

A flurry of similar cables flew out from Indian Harbor in the direction of other newspapers and news agencies before Peary left and hurried to Nova Scotia, traveling from there overland to Maine. En route he met up with two men, General Thomas Hubbard and Herbert Bridgman, both extremely wealthy and influential Peary patrons, and office holders of the Peary Arctic Club. The Peary Arctic Club was an informal panel of Peary advocates and supporters, the byproduct of Peary's talent for self-promotion, and it represented probably one of the most potent weapons in his arsenal.

The Peary Arctic Club was founded in 1898 for the simple purpose of supporting and funding Peary's various expeditions. By then, Peary had certainly proved himself among the elite of the exploration fraternity, and as such he had gathered around him a strong body of patrons and supporters. It has often been claimed that the rewards of membership of this club was the privilege of naming recently discovered geographic features, and although patrons did enjoy this honor, the Club certainly aimed at more than that.

The three men, en route back to the United States, immediately set about devising a strategy to undermine Cook's claim. One problem, however, was the fact that, even having disposed effectively of Cook's expedition records, and therefore any basis upon which he would have to support his claim, Peary's own expedition records were sporadic, ambiguous, and extremely thin.

Both parties had brought along a trained navigator, but both parties had turned their navigators back before reaching the North Pole. Furthermore, Peary's daily log records that the pace and daily distance traveled by the expedition almost doubled at the moment Peary was unencumbered by a second navigator. This marks the first troubling point of Peary's account - improbable distances appear in the record that would inevitably cast the overall credibility of his stated position in doubt.

On top of that, Peary's expedition recorded hundreds of miles of travel across featureless ice without any note of a pause to make celestial observations and establish longitude. This implies that the north/south axis upon which Peary would have traveled to direct himself towards the North Pole was entirely guesswork or some other system of spot navigation that he chose not to reveal. According to Matthew Henson, after five weeks of travel, Peary finally did make an observation, but when completed, he refused to share the result with his companions. He also left the pages of his diary for that particular day blank, and Henson further reported that he looked angry and disappointed afterwards. Later, he would claim that it was these observations that confirmed the arrival of the expedition at the North Pole.

In essence, what this implies is that Peary traveled 500 miles or so in a direction that he assumed was north and planted the flag at that point upon the assumption, or the "feeling," that it was the North Pole. His reckoning was backed up by very little observation, and though he was a surveyor whose guesswork could be expected to be more accurate than most, it was hardly enough to base a claim on.

To deal with this shortcoming, Peary, Hubbard, and Bridgeman devised a strategy. To buy time, Peary adopted the moral high ground by challenging Cook to prove his claim first. A statement was released by General Thomas Hubbard, on behalf of Peary and the Peary Arctic Club, stating simply, "Concerning Dr. Cook, let him submit his records and data to some competent authority, and let that authority draw its own conclusions from the notes and records...what proof Commander Peary has that Dr. Cook was not at the pole may be submitted later."

On September 21, 1909, the same day that Peary arrived in Nova Scotia, Cook made his grand entrance into New York, where he was greeted as a national hero. The streets were lined in celebration, and Cook delivered a public address, commencing with the triumphant words, "I have come from the North Pole." The next day, he met with reporters at the Waldorf Hotel in midtown Manhattan, and there he displayed the diary that he had kept with him. This, for the time being at least, was the only authentication that he had to offer. When reporters asked to see the sextant that he had used to determine his latitude at the North Pole, as well as other details and records, he replied that all of that was en route from Greenland in the care of Harry Whitney, and that it would all be delivered to him shortly.

A few days later, however, Whitney cabled Cook in New York and broke the news to him that his notes and equipment were still in Greenland because Peary had refused to allow them on board. Cook was completely broadsided by this, and the implications of it when he heard must have been heartbreaking. That is, of course, whether it was actually inconvenient for Cook to lose the records.

This, from the perspective of objective observers, is the most intriguing aspect of the controversy. According to Cook himself, and according to others who saw him, he was genuinely mortified to learn of the loss of his documentation and equipment, which implies that the cache of documents contained authentic proof of an authentic journey. However, there are less charitable observations that conclude Cook was relieved, as the loss of records would prvent him from having to explain his expedition. With his records lost, all that remained was his word, and though only a few contemporarie knew it at the time, his word was not bond.

That said, the weight of the evidence does not suggest Cook was happy to have the records lost. His initial claim was made with an unmistakable air of sincerity, and even after he learned that his records and equipment were as good as lost, he continued to speak publicly, give interviews, and maintain his position. He made use of his diary as a means of supporting his report, and although this was regarded as an important item of proof and a valuable document, it was clearly not enough on its own.

By then, with the weight of the Peary Arctic Club behind him, not to mention an army of supporters, Peary's claim was boosted. With the *New York Times* and the National Geographic Society both on his side, Peary certainly had some powerful allies. Cook's claim was dismissed by the *New York Times*, which ran a relentless campaign, as "the most astonishing imposture since the human race came on earth."

As this campaign took hold, Cook found himself wilting. He did not possess the sheer force of the personality Peary did, and rather than punch it out in the arena of public opinion, he opted increasingly to step back from the controversy and leave the field open to Peary. The endorsement of the National Geographic Society, the support of the *New York Times*, and the sheer weight of resources available from the Peary Arctic Club all made Peary the heavier hitter.

Then, quite suddenly, Cook's companion on his purported summit of Mount McKinley in 1906, Edward Barrill, reappeared. The Peary Arctic Club found occasion to query Cook's claim to the summit of Mount McKinley, and Barrill, who had entertained friends and family for years on tales of his and Cook's McKinley expedition, now sought to deny it. The Peary Arctic Club soon produced an affidavit signed by Barrill and notarized on October 4, 1909 that claimed he and Cook had never reached the summit. The document was published in the *New York Globe*, owned by General Thomas Hubbard, who made the fair and honest claim that if Cook lied about his ascent of McKinley, then he may very well have lied about reaching the North Pole. Barrill, of course, was compensated for his cooperation, and the Cook camp was quick to take note of this fact. A figure of between $5,000 and $10,000 has often been quoted, but whatever the true figure, it was worth more to Barrill than the value of an entertaining yarn.

Next came oral testimony from the two Inuit crewmembers who had accompanied Cook, with both claiming that they had been nowhere near the North Pole. Later, when cross-examined further, they admitted to not knowing what the North Pole was, or why anyone wanted to go there. The men were nonetheless quoted as stating that Cook's expedition had traveled only a few days north on the ice cap, and a map upon which they marked their route was given as evidence.[5]

Believing that Cook had been thoroughly discredited, the National Geographic Society established a three-man panel to examine Peary's expedition data. All three panellists were also Peary Arctic Club members, and one was a member of the same Coast and Geodetic Survey that Peary had served on. Needless to say, all three were Cook skeptics.

The examination took place on November 1, 1909, in the half-lit baggage room of a Washington, D.C. railway station, and there, a most cursory examination of the expedition records was carried out. Two days later, the National Geographic Committee met, examined the findings, and announced that Peary's claim to the discovery of the North Pole was indeed the only genuine one. Cook's claim, in the opinion of the most august and influential geographic society in the world, was dismissed as fraudulent.

For Cook, this was the straw that broke the camel's back. He canceled a lecture tour, citing laryngitis as the reason, and fled fled to Europe. A friendlier reception awaited him in Denmark, where news of his great success had first been received. During that visit, he had promised that he would present himself at Copenhagen University, and this he did. However, to his enormous surprise, when his diary was examined by university experts, who had also anticipated studying all of Cook's records, it was pronounced that Cook's claim to the North Pole was unproven. This, of course, was simply a statement of fact that Cook's claim could not be adequately proved in the absence of a complete record, but it was construed by the Peary camp as "disproved." This version of the University of Copenhagen report was circulated, and once again, loud voices on both sides of the Atlantic ridiculed Cook as a liar and a fraud. Cook had been among the founding members of two New York clubs promoting Arctic travel and exploration – the

---

[5] The map was the one deemed too close to the fictional route traced by author Jules Verne

Explorers Club and the Arctic Club of America. He even served the former as its second president, but he was expelled from both at the moment that the decision of the University of Copenhagen was made public.

The American minister to Denmark, Maurice Egan, a strong Cook supporter, admitted ruefully, "The decision of the Danish University is, of course, final, unless the matter should be reopened by the presentation of the material belonging to Cook, which Harry Whitney was compelled to leave..." Sadly, that was never to be. Cook's records, although searched for, were never found. Historians still debate whether they actually existed, whether Whitney really hid them, or even whether Peary managed to have them destroyed. The truth will never be known, but what is certain is that without them, Cook had no real hope of fighting back.

While public opinion tilted increasingly in the direction of Peary, Cook took a sabbatical lasting about a year in Europe, during which he wrote his book, *My Attainment of the North Pole*. Of course, that didn't mean others would leave him alone. Kicking him down when he was already out, on May 21, 1910, the *New York Times* ran an article with the following headline: "Cook Tried to Steal Parson's Life Work" The paper published a letter, written by the eminent zoologist Charles Haskins Townsend, which purported to expose more of Cook's discreditable activities, and the supporting article began in what was by then its customary style: "Dr. Frederick Cook, the Bushwick Avenue Explorer, who professed to have discovered the North Pole, and has since been declared one of the boldest fakers the world has ever known, is receiving the condemnation of explorers and scientists of this country and abroad for another alleged attempt at faking which has just become public."

Between 1897 and 1899, Cook had served as physician on board a Belgian Antarctic expedition commanded by Adrien de Gerlache, a Belgian naval officer and sometime Arctic explorer. Incidentally, another member was a young Roald Amundsen, with whom Cook would become lifelong friends. Amundsen, in fact, remained one of Cook's few long-time supporters and apologists. The expedition was conveyed aboard the ship *Belgica*, which called in at Tierra del Fuego on its way south. There Cook met and befriended a local Anglican missionary, Thomas Bridges, who was also a keen amateur linguist, and this proved to be a source of common ground between the two men. Cook spent a pleasant few weeks in the company of the older man, and he learned a great deal about local ethnography which was in part his role on the expedition. Bridges had been at work for decades among the local Ona and Yahgan peoples, already by then verging on extinction. His life's work to date was a 30,000-word Yahgan grammar and dictionary, yet unpublished, that fascinated Cook. Cook borrowed the manuscript, promising to return it, but within weeks Bridges died of stomach cancer, and the manuscript remained in Cook's hands.

Cook did not return the manuscript, and for decades it was forgotten, but Charles Townsend, author of the *New York Times* letter and a member of the Explorers Club, somehow heard about the manuscript. He was the first to point the finger at Cook for fraudulently attempting to publish the manuscript, somewhat revised, as his own work. By then, the manuscript was not with Cook but the Commission de la Belgica, the body responsible for the archives and findings of the Belgian Arctic Expedition. If efforts to publish the manuscript were indeed underway, those efforts could only have been through the Commission de la Belgica, with Cook possibly attempting to claim authorship as the ethnographer of the Expedition.

When the news broke, however, Cook was still somewhere in Europe, and the Commission de la Belgica immediately dropped the project like a hot potato. Townsend relentlessly championed the exposure of what he believed was a brazen attempt at fraud, and he petitioned the Commission de la Belgica for the return of the manuscript to Thomas Bridges' son, Lucas Bridges, who would later achieve fame and recognition as a memoirist and his father's biographer. In reply, the Commission de la Belgica admitted that publication of the manuscript had begun under Dr. Cook's name, but that it had been interrupted thanks to the fact that Cook was nowhere to be found to add editorial details. In the end, Lucas Bridges, Thomas Bridges' son, would regain possession of the manuscript, and it would be posthumously published in part in his father's name in 1930.

Needless to say, the whole affair did nothing whatsoever to rehabilitate the reputation of a man already suffering criticism. The truth of the matter, as with everything related to Cook, was enigmatic to say the very least, but the testimony of Explorer Club fellow Herbert Bridgeman, in hearings to consider Cook's expulsion, seems very damning: "Doctor Cook used to talk to me about his investigations in Tierra del Fuego. He always told me that he had made a special study of the Onas and had learned enough of them and their language to write a full report for the Belgican Commission. His claim of having mastered the language of the Onas and the Yahgans is ridiculous. The six weeks he spent there was not enough time to master the subject sufficiently to compile a grammar or dictionary."

Cook himself had absolutely nothing to say publicly about the affair, and the evidence does not paint a pretty picture. There can be no doubt that Cook did keep possession of the manuscript after Thomas Bridges' death, and long before the Mount McKinley or North Pole controversies, he certainly did attempt to insert the document into the published records of the Commission de la Belgica as his own work.

Still, things did not go all Peary's way. Having received the endorsement of the National Geographic Society, Peary submitted to an examination by a Naval Affairs Committee of the U.S. House of Representatives. He assumed this would give him official recognition by the federal government, but the U.S. Navy demanded a somewhat higher standard of proof, and there was much about what Peary had to offer that was questioned. Even the fact that the pages of his notes and journals were so clean prompted some suspicion, as one representative wondered how no grease had gotten on any of the pages.

Peary conducted himself competently throughout the hearings, and even though a great many unanswered questions remained, it was generally acknowledged that the Committee had no choice but to take him at his word. Representative Thomas S. Butler of Pennsylvania put it this way: "We have your word for it, your word and your proofs. To me, as a member of this committee, I accept your word, but your proofs I know nothing at all about."

The House Committee, of course, was in a quandary, simply because there was no possibility of examining Cook. Cook was nowhere to be found, and even if he could be found, he had no documentary evidence to offer. He had on his record a false claim to the summit of McKinley, and as a known liar, what chance did he stand in an official forum? He relied instead on his official account, *My Attainment of North Pole*, and what that could not achieve could not be helped.

In the end, the House Committee reluctantly endorsed Peary's claim over Cook's by a vote of 4 to 3, expressing "deep-rooted doubts." The bill that entered the House and Senate would not use the term "discovery," instead crediting Peary with an "Arctic exploration resulting in [his] reaching the North Pole." Peary was placed on the retired list of the Navy's Corps of Civil Engineers with the rank of rear admiral, and he was awarded an annual pension of $6,000.

After that, Peary quit the fight while he was ahead. He buried his data and documentation, and he never discussed the matter openly again for the rest of his life. In the court of public opinion, he was acknowledged as the one who discovered the North Pole, and even today, when the question is searched, the name Robert Peary often appears. He certainly achieved the fame that he so desired, and the accolades that came with it.

When Cook returned to the United States from Europe in 1911, some members of Congress tried to reopen the matter, and there certainly was a willingness and an interest in re-examining the record, but the start of World War I eclipsed an issue that had by then anyway ceased to be of immediate interest.

While Peary went into ostensible retirement, Cook went into the oil business in Wyoming and Texas, but the aura of dishonesty did not leave him there. In 1923, he was indicted and convicted on mail-fraud charges related to the pricing of stock in his company. He was sentenced to 14 years and nine months in Leavenworth. His oil leases made a fortune for someone else, and he died on August 5, 1940, at the age of 75.

Peary, on the other hand, retired to the town of Harpswell on the coast of Maine, where he remained active in various clubs and forums. He served twice as president of The Explorers Club, from 1909-1911 and from 1913-1916. He continued to receive honors from various domestic and overseas societies, and his recognition as the first to reach the North Pole was never seriously challenged again in his lifetime. He died with the rank of Admiral in Washington, D.C. on February 20, 1920 and was buried in Arlington National Cemetery.

The Aftermath and the Investigations

"The error was doubtless committed in good faith, but most authorities today agree that no one has an untainted claim to be first - neither Cook, nor Peary, nor the two Eskimos and the American black, Matthew Henson, who accompanied Peary." – *New York Times*, 1988

In 1985, the National Geographic Society commissioned British explorer Sir Wally Herbert to reexamine the Peary Archive, made available by the Peary family with additional documentation, in the hope of laying to rest any ongoing uncertainty. There certainly was no anticipation that anything other than a confirmation of Peary's claim would result - in fact, the new investigation was intended to debunk a 1984 television documentary broadcast by the British ITV channel that painted a picture of a hapless and naïve Cook being robbed of his laurels by the malevolent and scheming Peary.

Herbert, who in 1968 walked across the Arctic, was an ardent Peary admirer, and he anticipated an easy confirmation of Peary's record. However, as he began his investigation, he immediately began to run into problems. Herbert knew the terrain, not to mention the difficulties and the realities of the Arctic, and he concluded reasonably quickly that Peary's claim to have gone "Furthest North" simply did not add up. The main issue in Herbert's opinion was Peary's recorded mileage and distances; the results indicated in Peary's notes would have required an average distance of 77 miles a day, which could not have been possible.

Herbert recorded this and other findings in a National Geographic Article published in September 1988, but he also offered a much more detailed breakdown of his analysis of Peary's route in a book entitled *The Noose of Laurels*. An exhaustive list of inconsistencies is recorded in this book, which left Herbert extremely skeptical of Peary's claim. With the understanding that the burden of proof lies on the explorer, Sir Wally Herbert reluctantly concluded that Peary had not supplied this.

The Noose of Laurels had a similar view regarding Cook's claim. In the opinion of the esteemed polar explorer, neither Cook nor Peary arrived anywhere near the North Pole. Both were either mistaken or fraudulent, and given everything that was known about both men and the expeditions, Herbert believed both were frauds.

National Geographic Magazine published Herbert's findings, but it was not yet quite prepared to acknowledge the inadequacies of its original report. It might be remembered that the National Geographic Society and the *New York Times* were key sponsors of Peary's expedition, and both announced their acceptance of its results almost from the moment that Peary made the claim.

Meanwhile, the *New York Times* did run a correction on its 1909 article proclaiming Peary's success. The article, published on August 23, 1988, a month before the release of the National Geographic piece, was simply entitled *"A Correction."* After a brief summarization of the events of 1909, and an admission that Peary's claim was accepted based only on his reputation, the newspaper conceded, "In fact, modern explorers have been unable to confirm or duplicate the claimed course of either American. Cook failed to provide documented proof, and later died in prison after being convicted for mail fraud. As for Peary, an analysis being published in National Geographic corroborates what skeptics have long contended - that his logs are unreliable and his final push, averaging 71 miles a day, seems incredible." The conclusion, again, was that neither Cook nor Peary came within reach of the North Pole. No further comment was made about the integrity of either claim.

In the wake of Herbert's work, the National Geographic Society commissioned an independent report from the *Navigation Foundation*, a private nonprofit group run by U.S. Rear Admiral Thomas D. Davies, with a mission statement for the preservation and promotion of the art of navigation.

To the surprise of many, on December 12, 1989, the *New York Times* ran an analysis of the 230-page report, and it contained a surprising headline: "Peary Made It to the North Pole After All, Study Concludes." According to the article, "The new study, conducted by the Navigation Foundation, a professional navigation society, used new analytical methods to look at photographs, celestial sightings, ocean depth readings and other records made by the expedition to conclude that Peary's final camp was no more than five miles from the pole, not scores of miles away as some critics have charged."

Suddenly, the Peary camp was on top again, and his supporters, a surprisingly determined and dedicated fraternity, were jubilant. The report offered no insights into Cook's claim, but it now appeared that a broad consensus concluded he did not reach the North Pole. It was simply a question of whether Peary had.

82 years after the controversy began, Peary doubters and believers met face-to-face at a symposium held at the U.S. Naval Academy in Annapolis, Maryland. This gathering was described by the *Washington Post* as "an avocation by dedicated astronomers, navigators and geographical sleuths." The symposium was not a judicial proceeding, of course, but an informal jury of experts returned a reasonably unanimous verdict that Peary had pulled a fast one. Five experts addressed the forum after a careful examination of Peary's much-studied notes and photographs, and out of those five, three of them claimed overwhelming evidence existed that Peary did not reach the North Pole. Those three included Arctic explorers with undisputed claims to the pole, A fourth speaker said simply that there existed a method that Peary "could" have used to navigate, even though it is not the method Peary said he did use. A fifth panelist, recruited by the National Geographic Society to defend Peary's position, was of the opinion that photographs Peary said he took at the North Pole offered a 75% chance that the explorer came within 23 miles of the North Pole, and that was good enough. Summing the matter up, Charles Burroughs, chairman of the Washington chapter of the Explorers Club, said "the evidence was

pretty well lined up against Peary. The pro-Peary side just wasn't very convincing to me."

Much of the debate centered around the fact that Peary had made no longitudinal observations, which meant he could not have been expected to steer an accurate south/north bearing while achieving his stated miles a day. The *Navigation Foundation*, with a considerable basis of experience in navigations, claimed that Peary's steering technique was plausible because Roald Amundsen had used a similar system, relying mostly on a compass, to reach the South Pole, also without longitudinal observations. This was subsequently disproved by an amateur historian named Ted Heckathorn, who established that Amundsen had indeed taken careful longitudinal observations. Indeed, Peary himself took careful longitudinal observations on most of his expeditions, but, conveniently enough, not for his most historic one.

In 1993, the journal *Science* summed up the new evidence: "Skeptics, however, have long argued that it would have been almost impossible for Peary to have reached the North Pole as quickly as he claimed without longitudinal navigation. Now that Heckathorn has shown that Amundsen took such measurements, Peary's claims become all the more suspect."

The latest work done on the controversy came in 2005, when British explorer Tom Avery mounted an expedition to try and replicate Peary's times and distances using wooden sledges and Inuit dog teams. Setting out from precisely the same location as Peary – Cape Columbia on Ellesmere Island – the expedition covered the 475 miles some four hours faster than Peary and his team, but the longest average distance traveled over the final five days was 90 nautical miles, which is about 103 miles. This left a team of highly trained and fit men utterly exhausted, and yet Peary, who was much older, had no toes, and had a painful leg fracture, claimed 135 miles a day during the same period. In his later book, *To the Ends of the Earth*, Avery expressed the opinion that Peary could have reached the North Pole when he claimed, but that Avery personally doubted it. Fearing controversy, however, Avery would not submit the records of his expedition to scrutiny, asserting that his efforts would be "nit-picked" by a skeptical establishment apparently unable under any circumstances to reach a consensus on anything.

In the end, the single point most dwelt upon in examination of Peary's records and his account is the manner in which he conducted the navigations. Sir Wally Herbert's examination, which remained the most authoritative technical assessment, described Peary as navigating like a rank amateur, making errors that could only be described as astonishing.

In all of this, Cook tended to remain in the background, and there are now very few willing to argue on his behalf. In 1997, however, author and historian Robert M. Bryce published a book entitled *Cook and Peary – The Polar Controversy Resolved*. Bryce returned to the books that each contender wrote about his journey to the North Pole, and he found himself siding with Cook. His reasons were not particularly scientific; based on the scope of Cook's intellect, his poetic imagery, and his wide-ranging scientific mind, Bryce simply believed in Cook's integrity. "I knew the story couldn't be as simple as history had declared it to be. This man was much more than just a con man. And even if he was a con man, what drove him to do what he did? There had never been a full-length biography of Cook. So I set out to solve the mystery. And that turned into a biography of both Peary and Cook because their lives are inextricably intertwined."

However, after many years of research into Cook's private papers, and the unearthing of a previous undiscovered diary hidden in a museum of astronomy in Denmark, Bryce reached the same broad conclusion that neither Cook nor Peary came anywhere near the North Pole.

If neither man made it to the North Pole, who was the first to set foot there? By the time the exploration fraternity had gathered its wits and the thoughts of explorers once again turned to the North Pole, World War I was over, Robert Falcon Scott was long dead, and technology had advanced to the point that man-hauling and dog sleds were not the only means to reach the inaccessible poles.

As if there wasn't already enough controversy surrounding the expeditions to the North Pole, more arose in 1926 when Richard Byrd claimed to have flown over the North Pole in a Fokker Tri-Motor Airplane on May 9. The well-respected National Geographic Society initially verified his claim, only to have more information come out later to indicate that there were inconsistencies in Byrd's data. For instance, papers that were once hidden later came to light, including one note he wrote to the pilot that read, "We should be at the Pole now. Make a circle. I will take a picture. Then I want the sun. Radio that we have reached the pole and are now returning with one motor with bad oil leak but expect to make Spitzbergen." This called into question whether he actually made it over the North Pole or just wanted to claim he had. As the years passed, available information has tended to suggest that inaccurate navigation kept the two away from the North Pole, and their claim has since failed to carry much water.

On May 11, 1926, just a few days later, a small dirigible, the *Norge*, took off from Spitzbergen carrying the great Norwegian explorer Roald Amundsen, along with fellow explorer Lincoln Ellsworth, pilot Hjalmar Riiser-Larsen, Norwegian naval officer Oscar Wisting, and an Italian aircrew led by aeronautical engineer Umberto Nobile. Two days later, they landed in Alaska after having drifted over the top of the earth. While people were willing to level accusations against Peary, Cook, and Richard Evelyn Byrd, the legendary Amundsen had already made a claim to the South Pole, so doubting him was more than anyone was prepared to do. No doubts were ever expressed about Amundsen's claim, and thus, in the opinion of many, it was he who first visited both poles.

Amundsen

   For his part, Amundsen admitted he had trained all his life for the event: "I irretrievably decided to be an Arctic explorer. More than that, I began at once to fit myself for this career. ... At every opportunity of freedom from school, from November to April, I went out in the open, exploring the hills and mountains which rise in every direction around Oslo, increasing my skill in traversing ice and snow and hardening my muscles for the coming great adventure. ... At 18 I graduated from the college, and, in pursuance of my mother's ambition for me, entered the university, taking up the medical course. ... With enormous relief, I soon left the university, to throw myself wholeheartedly into the dream of my life. ... As soon as my army training was over, I undertook the next step in my preparation for Arctic exploration. By this time I had read all the books on the subject I could lay my hands on, and I had been struck by one fatal weakness common to many of the preceding Arctic expeditions. This was that the commanders of these expeditions had not always been ships' captains. They had almost invariably relied for the navigation of their vessels upon the services of experienced skippers. ... Always two factions developed—one comprising the commander and the scientific staff, the other comprising the captain and the crew. I was resolved, therefore, that I should never lead an expedition until I was prepared to remedy this defect."

Had World War II not intervened, there is a good chance that the first person to reach the North Pole would have made it there by the end of 1938. In May 1937, the Soviet Union sent a team of scientists to establish an ice station just 13 miles from the Pole, and four men lived there for nine months, conducting experiments and gathering data before being picked up by two ice breakers, the Taimyr and the Murman, on February 19, 1938. By then, their little station had drifted more than 1,000 miles south to the east coast of Greenland. Of course, World War II did intervene, and the Soviet Union, along with most of the rest of the world, had to turn its attention and resources to fighting rather than exploring.

In May 1945, as the war in Europe was drawing to a close, a Royal Air Force plane piloted by David Cecil McKinley became the first British plane to complete a flight over the North Pole.

To this point, nobody had completed the journey overland. The first to cross the ice and reach the North Pole, assuming neither Peary nor Cook did, was Ralph Plaisted, a Minnesotan salesman and outdoorsman who reached the North Pole by snowmobile in 1968. A year later, Herbert planted the Union Jack at the North Pole after a more traditional expedition using sledges and dogs, but his expedition was resupplied by air, something the early 20th century explorers did not have going for them.

The first wholly unsupported overland expedition by dogsled was successfully mounted in 1986 by Will Steger and Paul Schurke. In 1995, a member of that expedition, Canadian Richard Weber, as well as Russian Mikhail Malakhov successfully made an unsupported return journey.

Since then, there have been regular visits to the North Pole, even by motorcycle, but pundits are increasingly pessimistic that traveling over ice to the North Pole will be viable for long, as stretches of open water grow wider and deeper every year due to climate change.

Online Resources

Other books about explorers by Charles River Editors

Further Reading

Berton, Pierre (2001). The Arctic Grail. Anchor Canada (originally published 1988). ISBN

0-385-65845-1. OCLC 46661513.

Bryce, Robert M. (February 1997). Cook & Peary: the polar controversy, resolved. Mechanicsburg, PA: Stackpole Books. ISBN 0-689-12034-6. LCCN 96038215. LCC G635.C66 H86 1997.

Coe, Brian (2003) [First published 1988]. Kodak Cameras: The First Hundred Years. East Sussex: Hove Foto Books. ISBN 978-1-874707-37-0.

Davies, Thomas D. (2009). Robert E. Peary at the North Pole. Seattle: Starpath Publications. ISBN 0914025201.

Fleming, Fergus (September 27, 2001). Ninety degrees north: the quest for the North Pole. London: Granta Books. ISBN 1-86207-449-6. LCCN 2004426384. LCC G620.F54 2001.

Henderson, Bruce (2005). True North: Peary, Cook, and the Race to the Pole. W. W. Norton and Company. ISBN 0-393-32738-8. OCLC 63397177.

Herbert, Wally (July 1989). The noose of laurels: Robert E. Peary and the race to the North Pole. New York, NY: Atheneum. ISBN 0-689-12034-6. LCCN 89000090. LCC G635.P4 H4 1989.

Mills, William J. (2003). Exploring Polar Frontiers: A Historical Encyclopedia. ABC-CLIO. ISBN 1576074226.

Nuttall, Mark (2012). Encyclopedia of the Arctic. Routledge. ISBN 1579584365.

Rawlins, Dennis (1973). Peary at the North Pole: fact or fiction?. Washington: Robert B. Luce. ISBN 0-88331-042-2. LCCN 72097708. LCC G635.P4 R38.

Robinson, Michael (2006). The Coldest Crucible: Arctic Exploration and American Culture. Chicago: University of Chicago Press. ISBN 978-0-226-72184-2.

Schweikart, Larry. Polar Revisionism and the Peary Claim: The Diary of Robert E. Peary, The Historian, XLVIII, May 1986, pp. 341–58.

American History, Feb. 2013, Vol. 47 Issue 6, p. 33

Brendle, Anna. "Profile: African-American North Pole Explorer Matthew Henson." National Geographic News. National Geographic Society, 28 Oct. 2010. Web. 29 Nov. 2015

Dolan, Sean. Matthew Henson. New York: Chelsea Juniors, 1992. Print.

Johnson, Dolores. Onward. Washington D.C.: National Geographic, 1949. Print.

"On Top Of The World" American History 47.6 (2013): 33–41. History Reference Center. Web. 16 Dec. 2015

"Robert Peary." American History. ABC-CLIO, 2015. Web. 21 Oct. 2015.

Free Books by Charles River Editors

We have brand new titles available for free most days of the week. To see which of our titles are currently free, click on this link.

Discounted Books by Charles River Editors

We have titles at a discount price of just 99 cents everyday. To see which of our titles are currently 99 cents, click on this link.

Printed in Great Britain
by Amazon